D-Forms

Surprising New 3-D Forms From Flat Curved Shapes

John Sharp

tarquin publications

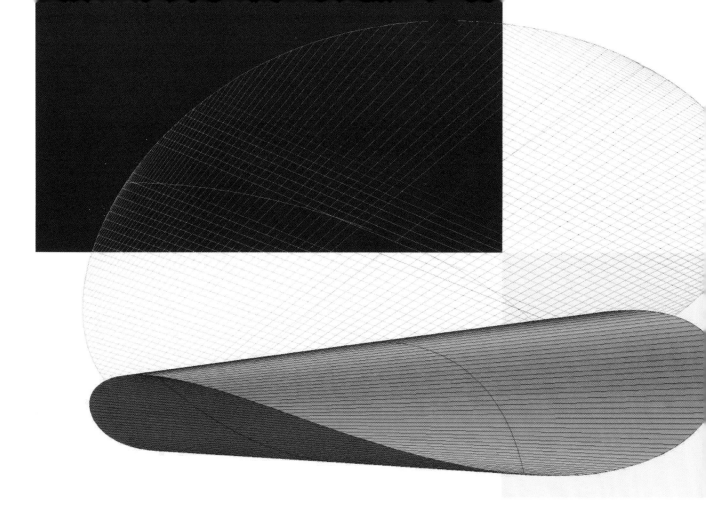

Designed and edited by Jeffrey Rutzky
Photographs by Tony Wills; concept renderings by Brian Watson
Additional photographs and illustrations by Jeffrey Rutzky, Brian Watson, Paul Bourke, and Kenneth Brakke
Tab design for Crescent packaging and Anti-D-Form photographs by Chris K. Palmer

Tony Wills looks forward to everyone developing ideas from D-Forms, and only requests you credit him with the original idea. Contact him at wills-watson.co.uk.

ISBN: 978-1-8996-1887-3

Printed in Malta by Melita Press

10 9 8 7 6 5 4 3 2 1

tarquin publications
99 Hatfield Road | St. Albans Hertfordshire AL1 4JL | England
+44(0)1727 833866 | tarquinbooks.com | info@tarquinbooks.com

Contents

D-Forms have been proposed for public installations, such as these formed in concrete by sculptor Tony Wills.

"There is no such thing as an ugly D-Form."—Tony Wills

What are D-Forms?

D-Forms are three dimensional forms created by joining the edges of two flat shapes having the same perimeter length.

It is surprising, given their simplicity of construction, that they have not been thought of before. The discoverer of D-Forms, Tony Wills, has opened up new areas of mathematics as well as many sculptural and design possibilities.

The models in this book are made from paper which is an ideal material for D-Forms. The original flat shapes of each D-Form should be made of a material that does not perceptibly stretch. Although you can get interesting effects with textiles, they are not strictly D-Forms.

D-Forms surprise us in many ways and you can only appreciate them by making the models. Depending on where you have chosen to start to join the two shapes, each face *informs* the other as to what three-dimensional form to finally produce. The emerging D-Form continually changes shape as the edge joining progresses. The final D-Form that results only appears when the process is complete and it is only by seeing this happen that you can understand the possibilities.

Creating D-Forms is unlike most forms of model making. Usually a set of model shapes can only be assembled in one way. However, a pair of shapes for making a D-Form can be joined edge-to-edge in an *infinite* number of ways.

Where Did Their Name Come From?

There are a number of reasons why they are called D-Forms. Tony Wills thought of them in a *d*ream; they are *d*evelopable surfaces and one of the earliest models had a cross section which is a *d*-shape.

Before You Start with the Models in this Book

Prior to cutting out the models in the book, you should explore the D-Forms concept first with a pair of identical elliptical shapes.

1. Fold a standard sheet of paper in half along its length, and unfold.
2. Trace the ellipse shown at the top of page 5 three times on half of the sheet.
3. Fold the paper back underneath your traced ellipses and cut out the shapes with scissors. You should have six ellipses to make three models.

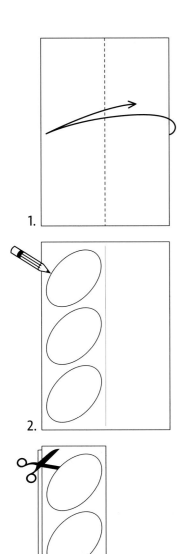

1.

2.

3.

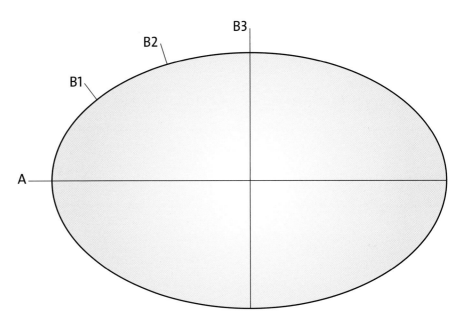

B3

B2

B1

A

4. Place each ellipse back down onto the template to mark as follows: Mark three of the ellipses at point **A**. Mark a fourth ellipse at point **B1**; a fifth at point **B2**; and the sixth at point **B3**.

5. Assemble the first pair of ellipses by matching up point **A** with point **B1** and secure with a thin strip of tape. Then carefully secure the remaining edges together, with thin strips of tape, along the entire perimeter. You now have a three-dimensional shape call a D-Form!

6. Repeat by matching up point **A** with point **B2**, securing as before. Note how the form gains volume.

7. Make a third D-Form by matching up point **A** with point **B3**, securing as before. The same ellipses are now offset 90°.

Geometrical Ideas

The following pages explore the geometry behind D-Forms and suggest ways you can create your own. Since D-Forms are a new mathematical discovery, their geometry has not yet been fully explored. This book is a first step towards this exploration.

Fresh ideas are occurring all the time as mathematicians, artists and designers see the possibilities. There is more on their artistic and design potential on pages 10 and 20–21.

The examples from the previous page show how the place chosen to start joining the two shapes leads to each face *informing* the other as to its final three-dimensional form. The only exception to this is when one of the flat shapes is a circle—no matter where you choose to join the two shapes together the result is the same.

What's Special About D-Forms as Surfaces?

Once they have been constructed as physical objects, it is possible to gain some insight into their properties, but it is not easy to predict the shape from the original shapes. Attempts are being made to create them in a computer system but there is no satisfactory method available as the book is being published. A method using approximation as soap bubbles illustrates the property of D-Forms as being unstressed; the shapes you are using for the construction move together as they are joined without any need to force them to do so.

D-Forms are a special type of surface called a *developable surface*, which means the surface can be cut open and flattened into a plane. They are also ruled surfaces, which means that the surface can be defined as the path of a straight line that moves across the surface. So a sphere can never be constructed as a D-Form because it is not possible to make a truly flat map of a sphere. (It is not a ruled surface and can only be mapped by projection not development of the surface; the gores used to make globes are only approximations.)

The edges of the pieces that are joined separate a D-Form into distinct surface areas. These areas may also have "invisible edges" where two developable surfaces blend into one another. The most obvious case of this is the **Squaricle** where you can see the part of the D-Form that arises from the square becomes a plane and four parts of cylinders at the corners.

Properties of Developable Surfaces

Developable surfaces fall into two main groups: cylinders and cones. There is a third type which are known as tangential developables *(see page 8)*. The surfaces of D-Forms can be parts of cylinders or cones or blended mixtures of the two.

Cylinders

Most people when asked what a cylinder is would describe a circular cylinder, but this is only one of many possible cylinders.

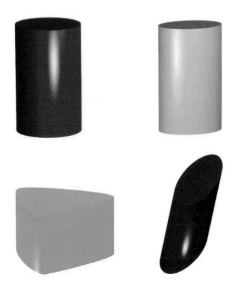

The simplest cylinder is the red one, and is known as a right circular cylinder. It is created by taking a circle in a plane and moving the circle at right angles to the plane to *extrude* the circle into a surface. Other cylinders are formed by taking a plane curve and extruding them along a straight line. The yellow cylinder is formed by taking an ellipse as the curve; the green a triangular shaped curve. The blue cylinder is formed by extruding a circle along a line which is not perpendicular to the plane of the circle. Because it is often necessary to compare two or more cylinders to see if they are the same, it is common to describe the cylinder by the curve formed by slicing the cylinder perpendicularly to the axis. This is known as a right cylinder (having a right angle between the axis and the end plane). When the curve defining the cylinder is infinite (like a parabola or hyperbola), then it is only possible to make a physical model of part of the cylinder.

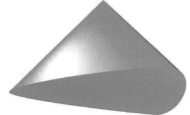

Cones

When it comes to cones, most people think of a right circular cone. There are many types of cones depending on the base curve of the cone.

Cones are formed by taking a plane curve and a point not in the plane (the vertex of the cone) and joining each point on the curve to the point. If the curve is a circle, when the line joining the point to the centre of the circle is perpendicular to the plane of the circle this gives a right circular cone as with the blue cone. The red cone is one where the base is an ellipse and the green one is a cone from a triangular curve.

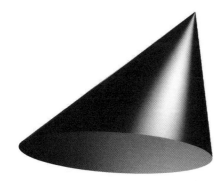

In constructing a cone, the perpendicular line from the vertex to the plane of the curve (the normal to the plane), can fall anywhere on the plane depending on how the plane is tilted. So when defining a cone it is necessary to define the plane curve and the relative position of the vertex in space. Varying the distance of the point from the plane also defines different cones for the same plane curve.

A geometrically defined cone extends either side of the vertex, since the line joining the point and curve can have any length. This makes it hard to create a true model of a mathematical cone and so models are normally only a small part of the cone. If a cone is transformed so that the vertex moves out to infinity, then the cone becomes a cylinder.

Ruled Surfaces

The extrusion of the curve along a line to form a cylinder causes all the points of the curve to be transformed into lines of the cylinder. Similarly, joining the points of the curve at the base of the cone to the vertex point yields a set of lines. For this reason, cones and cylinders are ruled surfaces.

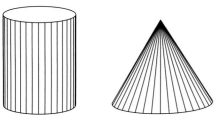

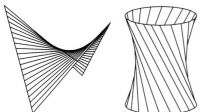

There are other types of ruled surfaces that are not developable. A ruled surface is not developable if it has a double curvature (for example having a saddle shape). The hyperbolic paraboloid is the most well known example of such a surface. The hyperboloid shape commonly seen in cooling towers of power stations is also a ruled surface which cannot be developed.

The difference between a ruled surface which is developable and one that is not is that adjacent ruled lines of a developable surface intersect (or are parallel in the same plane). The ruled lines of D-Forms are usually parallel and intersect the curves formed from joining the edges of the two shapes. This is described, and shown by experimenting with a ruler, in the **The Models Explained** on page 13.

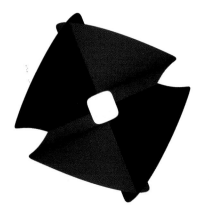

The D-Form and its curve.

Tangential Developables

The third type of developable surface is called a tangent developable because the ruled lines form the tangents to a curve. D-Forms do not include tangent developables as far as is known.

These surfaces are created by taking a curve in space (which does not lie in a plane) and drawing the tangents to the curve. This results in two surfaces that meet on the curve. Each half of the tangent gives rise to a developable surface; these half tangents form a clockwise and an anti-clockwise set. The curve on which they meet is called the edge of regression of the surfaces. (Cones and cylinders have an edge of regression which degenerates to a point, which is the vertex of the cone or the point at infinity on the axis of the cylinder.) The tangent lines form the generators that make the surface a ruled surface. The detailed description of developable surfaces requires a greater, complex book, so the following diagrams are only an introduction. The curve in these examples is the curve formed by the edge of a D-Form that arises from the intersection of two cylinders.

In these two views of one of the tangential developable surfaces you can see the curve in the center.

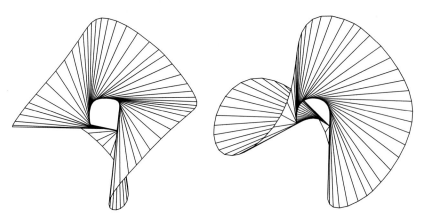

Various views of the two surfaces together are shown below. In this case they intersect (which may not always be the case) and both surfaces are the same but rotated in space relative to one another because the D-Form and the curve are symmetrical.

As the previous example shows, when you have created a D-Form there are two other associated developable surfaces for each edge. These are the tangent developables of the edge curve.

Exploring D-Forms, on page 20, has a section on Anti-D-Forms, which are created by joining the edges of two holes. These are not tangential developables, although they may look like them. No D-Form made from tangential developables is presently known.

Developing Surfaces

A surface is best developed by cutting along one of the ruled lines. When a cylinder (or cone) is defined by a curve that intersects itself, the surface of the cylinder goes through itself along one of the ruled lines; such a surface is hard to develop and may need to be cut into a number of pieces. When you develop a right circular cylinder by cutting along a ruled line, the result is a rectangle. When you develop a right circular cone by cutting along a ruled line, the result is a sector of a circle.

Elliptic cones give a variety of developments depending on how you cut the cone and the shape of the ellipse. The two following developments are of the same cone, cut at different positions.

D-Forms

The mathematics of developing known surfaces is usually easy to work with. However, the opposite process of taking a flat plane and creating a surface is not. Tony Wills' idea of creating D-Forms by joining the edges of two surfaces with the same perimeter seems very simple, and could have been studied at any time in the past 2,000 years. Yet nobody seems to have thought of it before him! It is almost impossible to make an ugly D-Form, so it is easy to create a multitude of beautiful sculptures which display an artistic side of mathematics never seen before. If one of the shapes is a circle, the other shape can only be joined in one way because of the circle's symmetry. In other cases, a whole range D-Forms can be constructed by varying the starting positions on the two shapes.

Tony Wills with stainless steel D-Forms.

Artistic and Design Possibilities

The geometry of D-Forms on the preceding pages is only one aspect. They also suggest ideas for artists and designers and it is no surprise that Tony Wills who discovered the concept of D-Forms is an product designer.

As the models in this book show, they have much in common with the sculptural forms of artists such as Barbara Hepworth and Constantin Brancusi who use surfaces meeting at a hard edge. The sculptors Naum Gabo and his brother Antione Pevsner have used developable surfaces and some of their work hints at D-Forms too.

As a product designer Tony has developed such products as the D-Form street furniture range which uses D-Forms as moulds into which is cast artificial stone to create elegant architectural elements *(see page 20)*. He is applying the sculptural aspects of the D-Form geometry to a variety of products. D-Forms have been investigated for aircraft propeller shapes, but their practical use is still waiting to be fully exploited.

Textiles

Textile designers use concepts similar to D-Forms to mould and shape garments, for example with gussets. Fabric does not have the same properties as paper or plastic, so the results are different.

One possibility for using textiles is to make cushions using the stuffing to create the surface by tensioning the fabric.

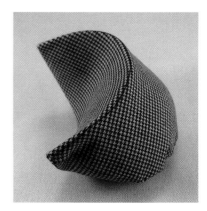

Applied D-Form by Tony Wills. Luminaire lamp made from Melinex polycarbonate with glass spheres.

D-Form Pillow Studies made from wool by Jeffrey Rutzky.

D-Forms | *Surprising New 3-D Forms From Flat Curved Shapes*

Making D-Form Models from Paper

How to Use the Model Pages

The principle characteristic of D-Forms is that different joins between the same pieces shape the final model. The models in the book have been designed to show different ways to create D-Forms. Each one has a specific method of joining. Photocopy the model pages before you cut them out so you can experiment with joining in different ways. The models are joined edge to edge with interlocking tabs using glue as shown below.

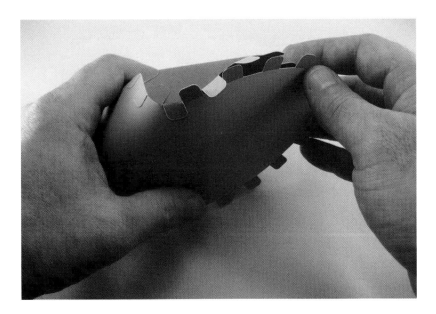

Use the templates provided at the end of the book to explore your own D-Form shapes. Fix the edges together with short lengths of tape. This method gives you more control over where you position the two shapes relative to one another. The tabbed method for the models in this book restricts you to a limited number of positions. Some models, however, like the **Squaricle** and **Wobbler**, can only be joined at the marks shown to obtain the desired result.

Assembly Instructions

Please read all the steps before you start.

1. Before starting to cut out, score the curve of each shape at the edge of the tabs. This makes the paper fold accurately and cleanly along the edge of the shape. Use a stylus or empty ballpoint pen to draw firmly along the boundary line. Two of the models (the **Squaricle** and **Crescent**) have creases that you need to score in the same way.

2. Cut out the models using a craft knife or sharp scissors. Take care to cut the edge that is in-between the tabs and not cut off the tabs themselves.

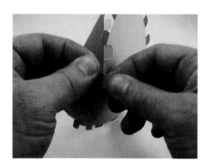

3. Find the corresponding dots on the two shapes where you should begin to fit them adjacent to one another.

4. Use a toothpick or a partially straightened paper clip to spread a thin layer of glue on the back of a tab. Use a glue which sets quickly, but not instantly. Work slowly around each perimeter, gluing only a few tabs at a time, and allowing them to set before proceeding. PVA craft glue is ideal because it dries clear if any glue comes out from underneath the tabs. Alternatively, a glue stick will also work well.

5. Work around the edge of the shapes. Use a mixture of pushing and pulling the tabs between one another to get them to mesh together. Pull *gently* on an adjacent pair of tabs until they lie along the resulting curve.

6. The shape of the D-Form grows as you join more of the edge. So make sure that the folded edges of the tabs of one shape align with the edge of the other one. This is harder to do initially, but gets easier as the final form settles down.

7. You might find that a model sometimes develops a dent. *Gently* push the edges of the model and you will find it pops out. When making the models, you might also find that pushing a pencil inside the model, before it is complete, allows you to force out any major dents.

There are some additional instructions with some models that apply only to those models. These appear on the model pages.

A wide variety of D-Form models you can make with this book.

The Models Explained

Model 1 | Twin-Ellipse

This model joins a pair of ellipses symmetrically.

An ellipse is a circle that has been squashed in one direction. The amount of squashing is measured by the ratio of the long (major) axis to the short (minor) axis.

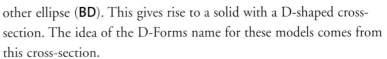

To make a **Twin-Ellipse**, join the ends of the major axis of each ellipse (**AC**) to the ends of the minor axis of the other ellipse (**BD**). This gives rise to a solid with a D-shaped cross-section. The idea of the D-Forms name for these models comes from this cross-section.

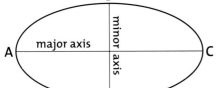

As the original ellipses were identical, the two curved surfaces are identical and are both convex.

If you place the model on a flat surface, and roll it on the surface, you will see that it will always be in contact with the surface along a line. This is not surprising since the surfaces are developable and developable surfaces are always ruled surfaces. If you move the edge of a ruler over one of the surfaces, you will see that the ruled lines are parallel to the minor axis of the ellipse. This means that the two surfaces are parts of cylinders.

Model 2 | Twisted Twin-Ellipse

This model uses the same ellipses as in the first one, but they are joined using a different starting point. Instead of getting a D-shape, the surfaces are twisted. They are also identical surfaces because the ellipses are the same. Suppose you were to join the ellipses so that the edges could move along one another. The initial flat shape you get when the ellipses match would give rise to a twisted shape which would then get less twisted as it got closer to the D shape.

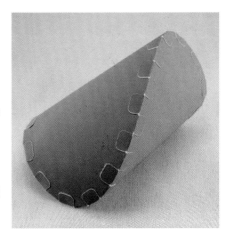

Even though it has a twist, if you place the model on a flat surface, and roll it on the surface, you will see that it will always be in contact with the surface along a line. The lines will be parallel showing that the surface is a cylindrical ruled surface. You can also see that the lines are parallel if you move the edge of a ruler over one of the surfaces. Unlike with Model 1, the lines are at an angle to the axes of the ellipse.

Model 3 | Twisted Duo-Ellipse

This model advances Model 2 a step further by using two different ellipses with the same perimeter. It is offset similarly to Model 2, but with the lengths of the axes being different, the longer, thinner, ellipse appears less twisted.

Rolling on the surface, and using a ruler to see the parallel lines of the defining cylinders, applies as with the previous models. Place the model on a flat surface with the ellipse which is closest to a circle on the surface, and rock it slightly. Let it come to rest. Do this again a few more times and you will see that it has two preferred equilibrium positions. If you do this with the longer ellipse side, there is only one.

Model 4 | Circlipse

When one of the shapes is a circle there is only one possible D-Form.

A circle is a special case of an ellipse, with an infinite number of axes of symmetry through the centre. This means that no matter where you start joining the perimeters to create the D-Form, you will always get the same result. Cross sections through the major and minor axes of the ellipse will always yield a D shape. So the model is convex in both directions and will rock if displaced from its resting position. The frequency is higher if the ellipse surface is being rocked.

If you move a ruler over the part of the surface created by the ellipse, the ruled lines are parallel to the minor axis of the ellipse and over the part created from the circle, they are parallel to the major axis of the ellipse.

Model 5 | Super Circlipse

This model results when the edge of a "super-circle" is joined to an ellipse.

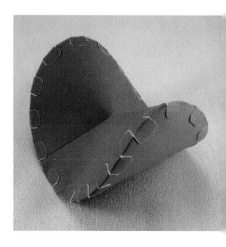

The development of a right circular cone results in a segment of a circle, whose height determines the angle of the segment, which is always less than 360°, becoming closer to it as the height diminishes and so being 360°

when the height is zero. What happens if you go past 360° and you create a "super-circle"? This is equivalent to joining two segments of circles. The images at left show how the resultant "cone" looks two segments of 220° are joined :

This model joins the edge of a super-circle to an ellipse. The ellipse takes up a D-shape, but the centre of the super-circle is now below the line joining the ends of the major axis of the ellipse. The D-Form is now a mixture of convex and concave parts.

If you move a ruler over the part of the surface created by the ellipse, the ruled lines are parallel to the minor axis of the ellipse as in Model 4. Moving it over the super-circle part, you can see that it is made from parts of four cones tangential to one another. At the ends next to the major axis of the ellipse they are convex, whereas next to the minor axis they are concave. At the transitional point they have a common tangent plane, so the ruler moves smoothly over the surface.

Model 6 | Tony Wills' Squaricle

When Tony Wills had discovered D-Forms, he had the idea of joining a square to a circle. As shown on the title page, he had a version of this made in gold-plated bronze.

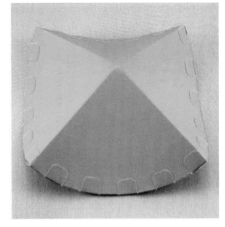

For a circle of unit radius, the square has a side of $\pi \div 2$ or 1.5707.

You might like to try drawing a square and circle and joining them using tape. It's quite hard to do since, as you join the edges, the paper crumples. Tony had the idea of creasing the circle along two perpendicular diameters. This gives a very interesting D-Form which he named the **Squaricle**.

The model is coloured to show the different developable surfaces of which it is composed. The circle becomes four sections of cones. The square becomes a plane (in the shape of a square whose vertices are the midpoints of the original square) and four circular cylinders.

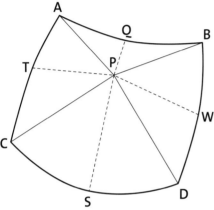

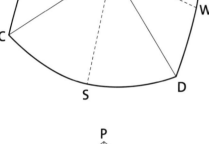

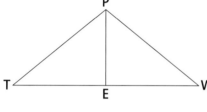

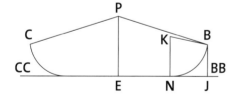

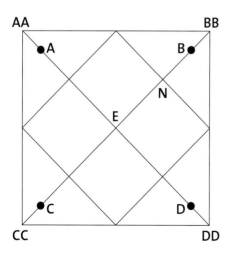

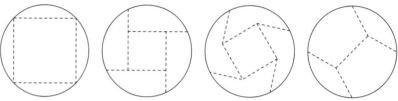

Depending on how the circle is scored, a wide variety of Squaricle forms can be made as shown above.

If you move a ruler over the base of the model, you can see the transition from the plane of the square to the cylinders. The sections of cones are obvious. This allows a geometric analysis of the **Squaricle**. The drawings show the distinctive cross sections and the positions of centres of the cylinders which then allows a computer model to be created.

The top figure labels the points on the standard **Squaricle**. The next one shows the cross-section through the mid-points of the square and below this along the crease lines. The fourth figure shows the position of the corners when they are folded up. In the third figure, you can see the position of the axis (**K**) of the cylinders (arc **NB**).

The **Exploring D-Forms** section on page 20 has some more ideas for different creases and creating D-Forms using other polygons. You might also think about models using the super-circle instead of a circle.

Try using different materials to join a square and a circle. This leather pillow (which does not crumple like paper) has been sewn together without creasing the circle.

D-Forms | *Surprising New 3-D Forms From Flat Curved Shapes*

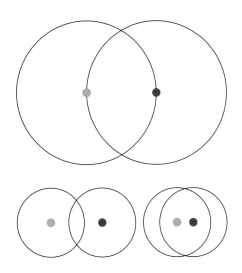

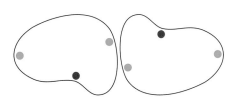

Model 7 | Crescent

This model is made from a pair of over-lapping circles, a device used in many forms of art which is known as the Vesica Piscis, or vessel of fishes.

The circles have the same radii and are drawn so that the centre of each is on the circumference of the other.

This is not strictly a D-Form when starting with the three shapes joined, since it is one piece. As with the **Squaricle** model, it is a D-Form with creases. You could, of course, have created it from three separate shapes. This suggests a whole new unexplored area of D-Forms where a set of shapes with common length perimeters are joined together.

You might also like to experiment with other circles where the distance of the centres varies. How far can you separate them before the edges will no longer join up?

If you follow the ruled lines in the surface by moving a ruler over it, you can see that the area of overlap of the original circles forms a cylinder (but not a circular one) and the sloping surfaces are parts of cones.

Model 8 | Twisted Bean

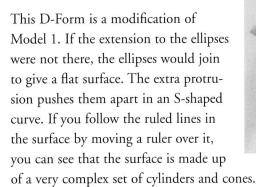

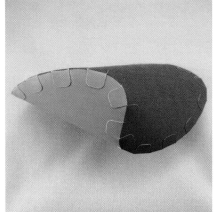

The shapes for this model are created from an ellipse with an added S-shape curve.

This D-Form is a modification of Model 1. If the extension to the ellipses were not there, the ellipses would join to give a flat surface. The extra protrusion pushes them apart in an S-shaped curve. If you follow the ruled lines in the surface by moving a ruler over it, you can see that the surface is made up of a very complex set of cylinders and cones.

Model 9 | Twisted Twin Cones

This model is made from an identical pair of elliptical cones. The ellipses on the bases of the cones are joined in the manner of Model 2.

If you take a cone which is not a right circular one, for example one which has an elliptical base, then you can join the edges in the same way as making the D-Forms using the ellipses in models 1 to 4. Making such a 3-D form requires the cones to be flexible, so they have to be constructed from a development of the cones you are working with and not a pair of solid cones. Model 9 is a pair of elliptical cones whose bases are joined. As with joining a pair of ellipses to make a D-Form in models 1 to 4, this can give rise to many different D-Forms. The 3-D curve formed from the edges of the original ellipses of the cone bases is particularly beautiful and much more prominent that the D-Form formed from two flat ellipses.

When you have made the model, take a ruler and place it on the surface and see what a complex set of developable surfaces it is composed of, each flowing into the next. Mostly the ruled lines are parts of cones which have the common apex of the original elliptic cones.

Model 10 | The Wobbler

The name for this model comes from the apparent movement it makes when rolling down a slope.

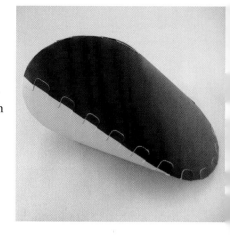

If you take this model and place it on a slightly inclined surface it will roll down the surface (you might need to give it a slight push to start it) and will keep rolling in a straight line, that is to say the ends move along two parallel lines. However, because of its shape, it appears to be wobbling.

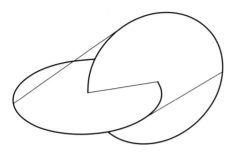

One form of the **Wobbler** consists of a pair of similar circular disks which are slotted together so that their centres are √2 times the radius of the discs. You could make this version by cutting radial slots in a pair of CDs. (You could also experiment with other separations of the centres and see how this affects the rolling.) The Wobbler surface is formed as the set of ruled lines which join the points on the two discs as they rest on a plane in various positions.

When you have made the Wobbler model, take a ruler and place it on the surface and see where the ruled lines occur. Look at it when it is placed on a flat surface and roll it gently to see how it is always in contact with the surface along a line.

You could try making other versions by taking the two pieces with the central part connected and then join the remaining edges with a different starting point. The model as designed is probably the only instance where it will roll and wobble.

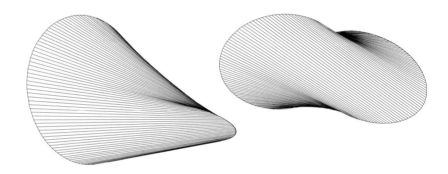

Above: *George Hart's Two-Cent Toy.*
Right: *The Wobbler appears to wobble when rolling, but it is always in contact with the surface along a line.*

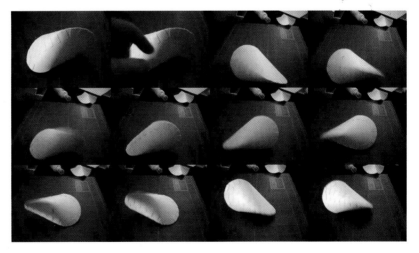

Exploring D-Forms

D-Forming is a new way of creating surfaces, so there is a great deal of scope for discovering new forms. Just describing the models in this book and their properties gave rise to some new ideas. The fastest way to make D-Forms is to join the edges together by using small pieces of tape the width of the tabs in the models in the book, or smaller, so that the edges of the two shapes join easily, and are not distorted. The use of a heavy but small tape dispenser is highly recommended. The templates on pages 43–44 will get you started, and then you can draw your own curved shapes. For complex curves you will have to join the same shaped pieces, unless you have access to computer software which can measure the length of the curves. The perimeter of some curves and polygons are easy to calculate, as in the examples below. You can also develop the combination of creases and D-Forms as Tony Wills has done with his **Squaricle**.

Ellipses

There is no simple exact formula for calculating the perimeter of an ellipse. However boat builders use a formula that is good enough making your own models.

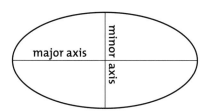

Perimeter of an ellipse = $(1.5(a + b) - ab) \div \pi$
a = major axis; b = minor axis

Applied D-Forms by Wills Watson+Associates. **Above:** *Stainless steel public sculpture.* **Right:** *Pre-cast stone street furniture made in fibreglass moulds.*

Variations on the Squaricle

The first, obvious variation is to create
D-Forms with other regular polygons.
You need to crease the circle with the
number of creases equal to the vertices
of the polygon. You may need to draw
the polygon by dividing a circle using
a protractor or on the computer since
it is not possible to draw all regular
polygons with a ruler and compasses.
If the circumcircle of a polygon with n
sides has a radius r, then the perimeter
of the polygon is $2r \sin(180° \div n)$. You

could also use polygons that are irregular, but this requires calculating
each side in order to find the perimeter and so determine the circle
with the same radius.

You could also use the ellipse calculation above and use an ellipse
instead of a circle. If you use an ellipse, because it only has two axes
of symmetry (and not an infinite number like a circle), there are
theoretically many ways to make a "Squellipse" from an ellipse and
a square. However, without the use of a computer program, it is not
possible to divide the perimeter of an ellipse into equal parts. You can,
however, crease along the axes.

Another variation would be to crease the circle in different ways as
shown on page 16. Provided that the creases end at the ends of a pair
of a perpendicular diameters, there are many possibilities of which the
following are only a few. You can even vary these by just creasing the
segments near the circle.

This one which has a flat base suggests architectural
forms.

Also keeping the creases end at the ends of a pair
of a perpendicular diameters, you can move the
centre. This (and the other possibilities) is also applicable to the ellipse.

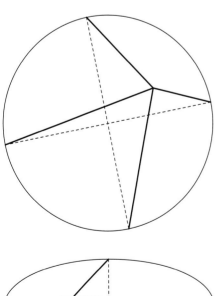

You will find that with some of these the crease lines do not remain
straight. So try deliberately making the lines curved while ensuring
that they start and end where you intend them to.

If you divide the circle by rotating one diameter, then you can join
the edge to a rectangle instead of a square. The ratio of the sides of
the rectangle is the ratio of the two angles between the lines. So if they
are inclined at an angle of 60°/120°, you have a 1:2 rectangle. If the
shortest side is one unit, the perimeter is six units and the radius of
the circle you require is $3 \div \pi$ or 0.955.

Anti-D-Forms

Another possibility which generates many more D-Forms is if you cut holes in piece of paper and join the edges of the holes. These Anti-D-Forms are just one possibility. You could also join the edge of a hole to the edge of a shape, such as cutting an elliptical hole and rotating the ellipse you have cut out before joining it again. This is called a Demi D-Form. There are no models of Anti-D-Forms in this book because ideally you need a large piece of material in which to cut the hole.

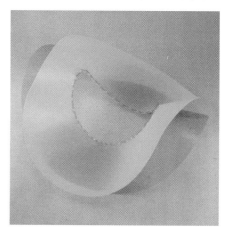

Dynamic Anti-D-Form made from polypropylene by Jeffrey Rutzky.

It might appear that these Anti-D-Forms are tangential developables. This is not the case. When you make a model, the shape of the surface with the hole is partly defined by the properties of the material it is made from, how big it is, and how it behaves under the force of gravity. Geometrically it follows the same principles as for the other D-Forms. For example, suppose you made a version of the **Circlipse**, with the ellipse replaced by a hole. The circle will form the same shaped cylinder. The surface around the elliptical hole would be a continuation of the same cylinder that forms the cylinder from the ellipse in Model 4. This gives another possibility. Instead of cutting a hole, cut slots which match the tabs of the ellipse and push the tabs of the circle through them.

Using Different Materials

Stiff paper is the ideal material for making D-Forms, but there is no reason why you should not use other materials. Some of the examples on page 10 use metal, plastic, and fabric. There is also a version of the **Squaricle** on page 16 sewn from leather. Ideally, the material should be very thin. Fabric has a tendency to stretch in directions along the warp and weft, but this should not be a problem. If you use filling to stuff a cushion, then the developable surface is deformed, and a totally new set of D-Forms can result. Using a stiffening material sandwiched between two sheets of fabric can preserve the developable surface, such as the **Crescent** pillow, shown on page 10.

Applied D-Forms by Jeffrey Rutzky. Above: Crescent packaging made from polypropylene. Right: Circlipse pillow made from polyurethane.

Using the Templates on Pages 43–44

Enlarge them on a photocopier and explore the possibilities for creating a multitude of D-Forms using the sticky tape method of joining as described on pages 4–5.

Model 1 | Twin-Ellipse

Cut out the shapes. Match the corresponding white dots on the reverse side to find the starting positions for creating the D-Form.

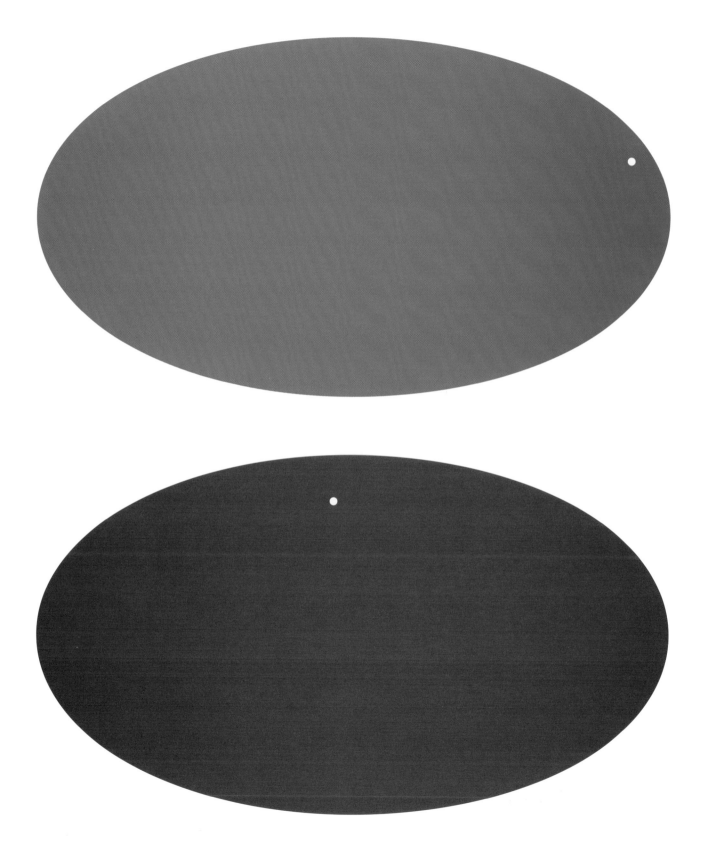

D-Forms | *Surprising New 3-D Forms From Flat Curved Shapes*

Model 2 | Twisted Twin-Ellipse

Cut out the shapes. Match the corresponding white dots on the reverse side to find the starting positions for creating the D-Form.

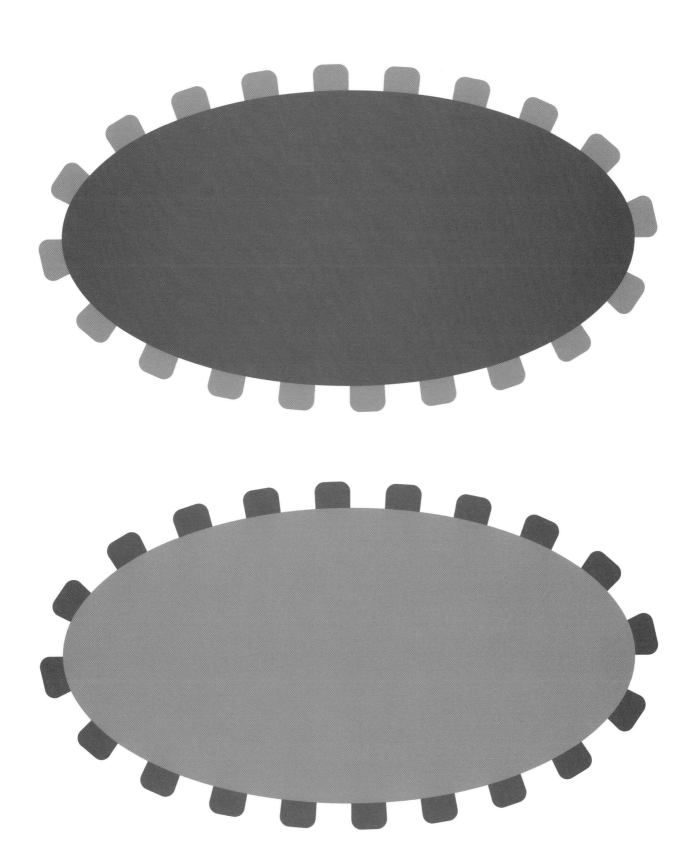

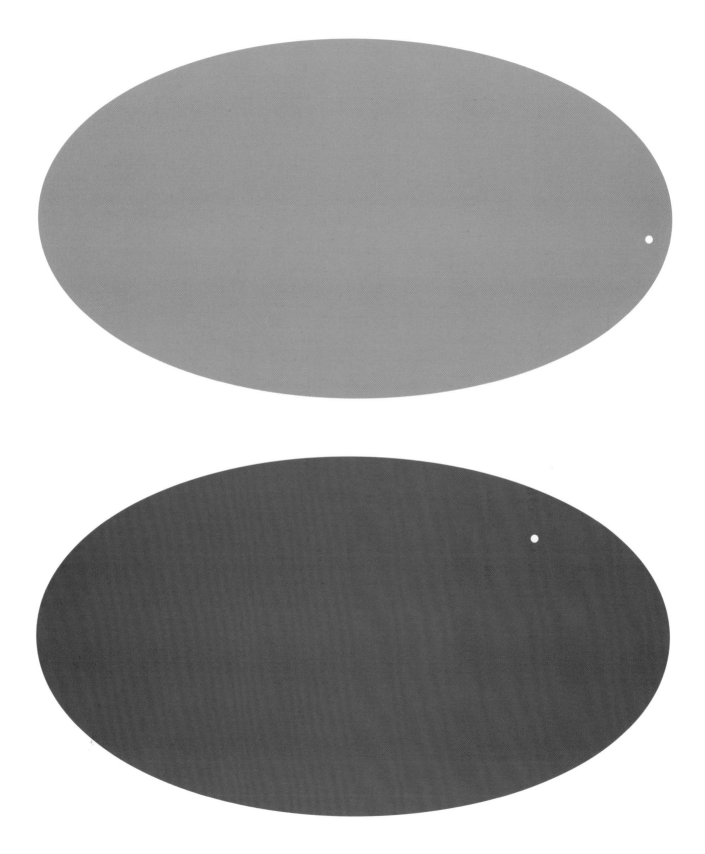

D-Forms | *Surprising New 3-D Forms From Flat Curved Shapes*

Model 3 | Twisted Duo-Ellipse

Cut out the shapes. Match the corresponding white dots on the reverse side to find the starting positions for creating the D-Form.

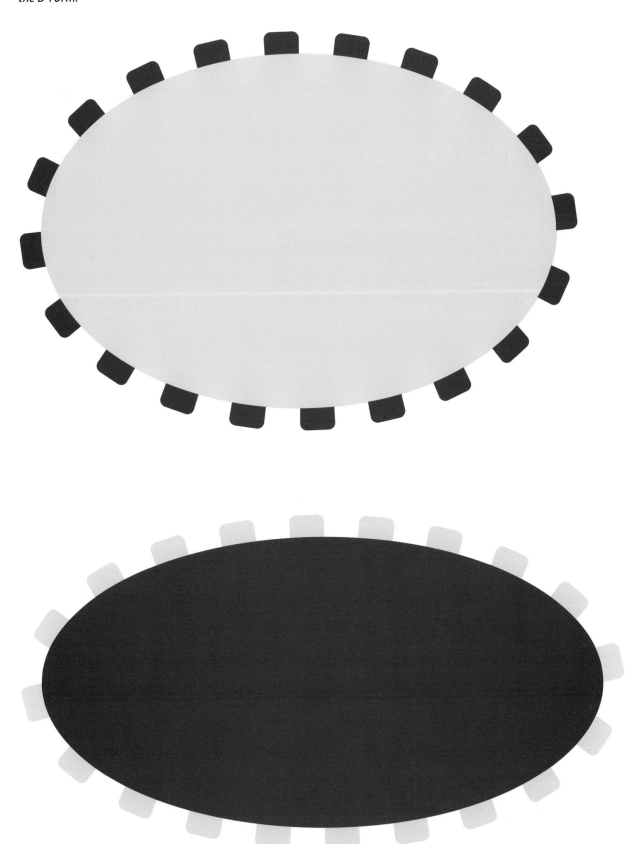

D-Forms | *Surprising New 3-D Forms From Flat Curved Shapes*

Model 4 | Circlipse

D-Forms | *Surprising New 3-D Forms From Flat Curved Shapes*

Model 5 | Super Circlipse

Cut out the shapes. First create the super-circle by sticking tab A to the edge marked A on the reverse side of the second semi-circle. Repeat with tab B. Then match the corresponding white dots on the reverse side to find the starting positions for creating the D-Form.

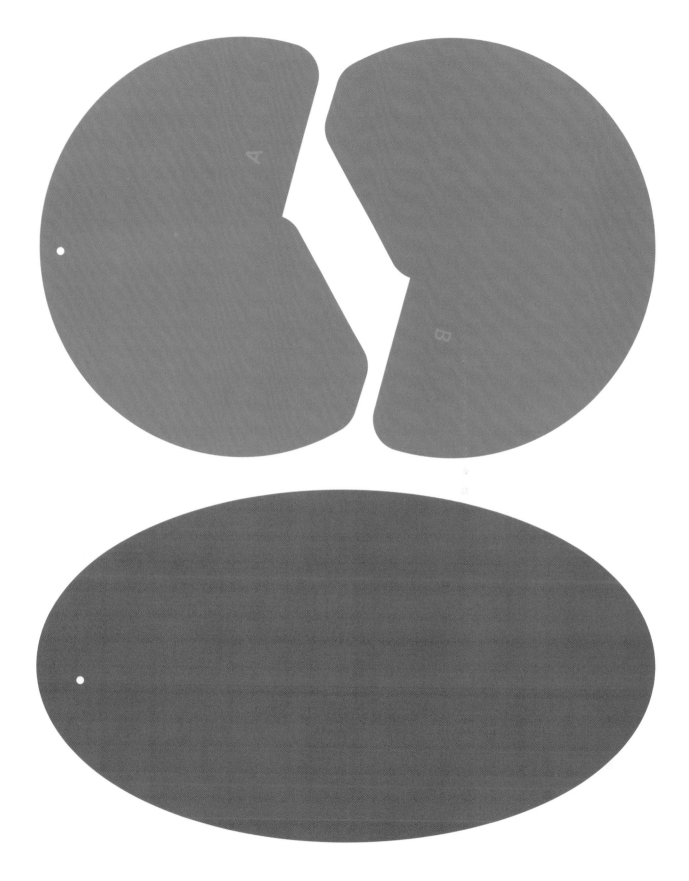

D-Forms | *Surprising New 3-D Forms From Flat Curved Shapes*

Model 6 | Tony Wills' Squaricle

Cut out the shapes. Use a stylus or empty ballpoint pen to draw firmly along crease lines. Match the corresponding white dots on the reverse side to find the starting positions for creating the D-Form.

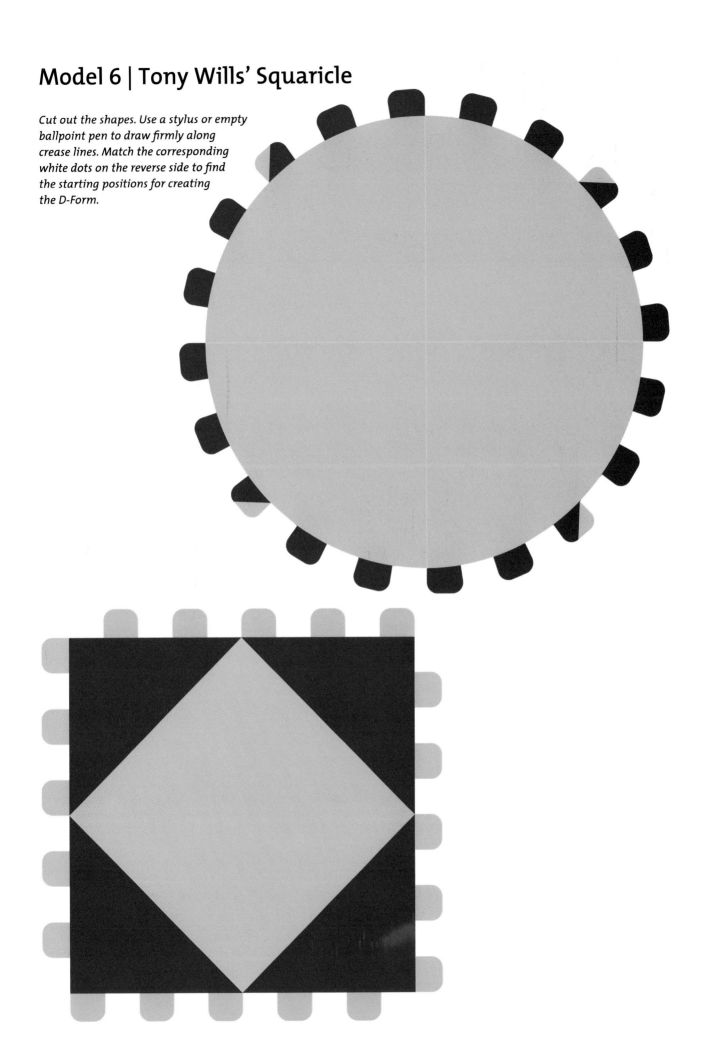

D-Forms | *Surprising New 3-D Forms From Flat Curved Shapes*

Model 7 | Crescent

Cut out the shape. Use a stylus or empty ballpoint pen to draw firmly along the edges of the yellow shape. Match the corresponding white dots on the reverse side to begin shaping the Crescent.

D-Forms | *Surprising New 3-D Forms From Flat Curved Shapes*

Model 8 | Twisted Bean

Cut out the shapes. Match the corresponding white dots on the reverse side to find the starting positions for creating the D-Form.

D-Forms | *Surprising New 3-D Forms From Flat Curved Shapes*

Model 9 | Twisted Twin Cones

Cut out the shapes. First create each cone by sticking the tab to its own same-colored edge. Then match the corresponding white dots on the reverse side to find the starting positions for creating the D-Form.

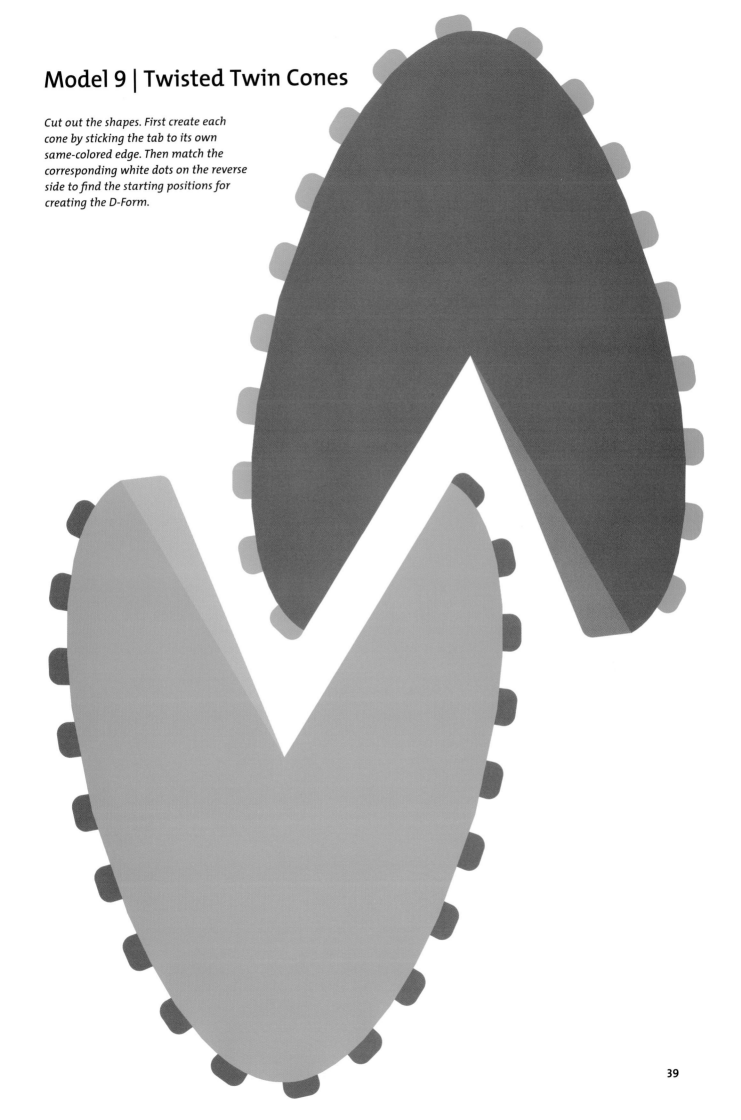

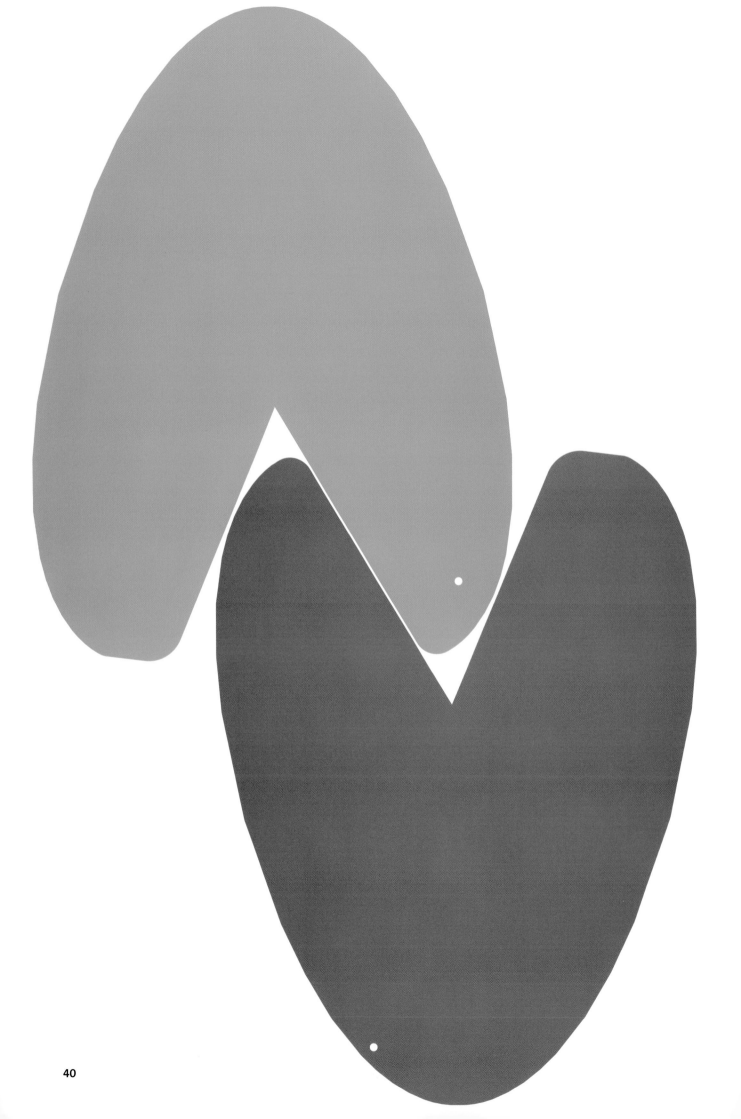

Model 10 | The Wobbler

Cut out the shapes. Start by sticking the four pairs of tabs together on the V-shaped section of each piece. Then match the corresponding white dots on the reverse sides to find the starting positions for creating the D-Form.

D-Forms | *Surprising New 3-D Forms From Flat Curved Surfaces*

Templates to Make Other Models

These templates are sets of shapes which all share the same perimeter. Use the marker line, which is one-tenth of the perimeter of the shapes, as a guide to make others, such as a square.

1/10 Perimeter

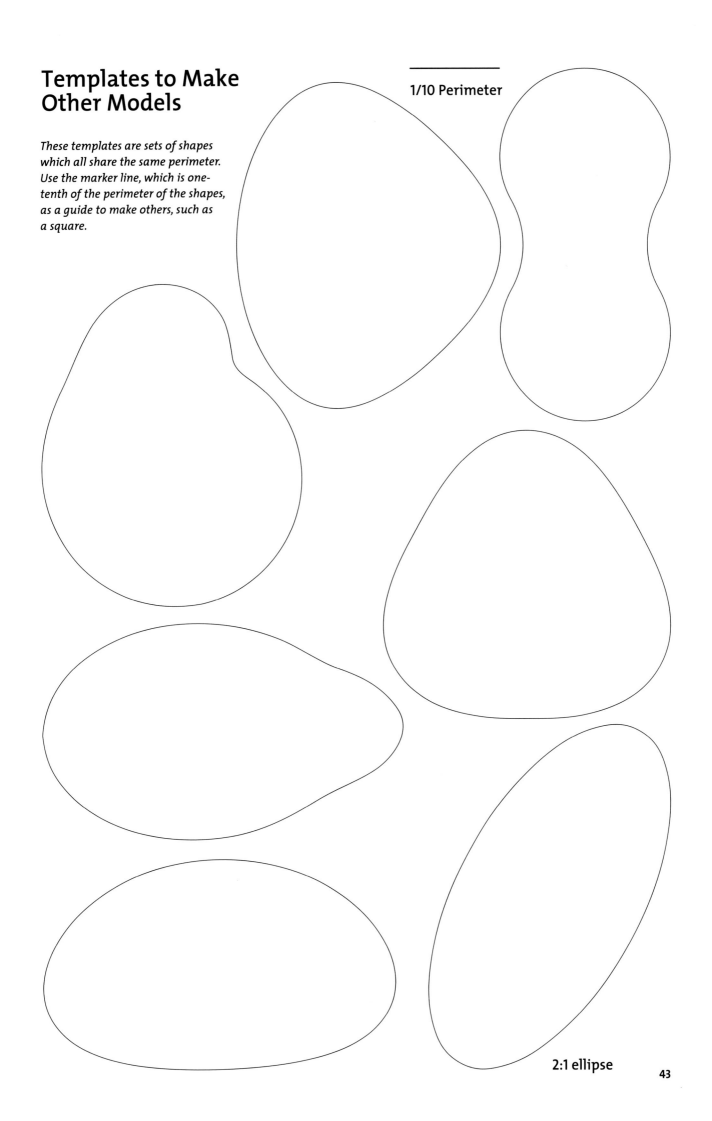

2:1 ellipse

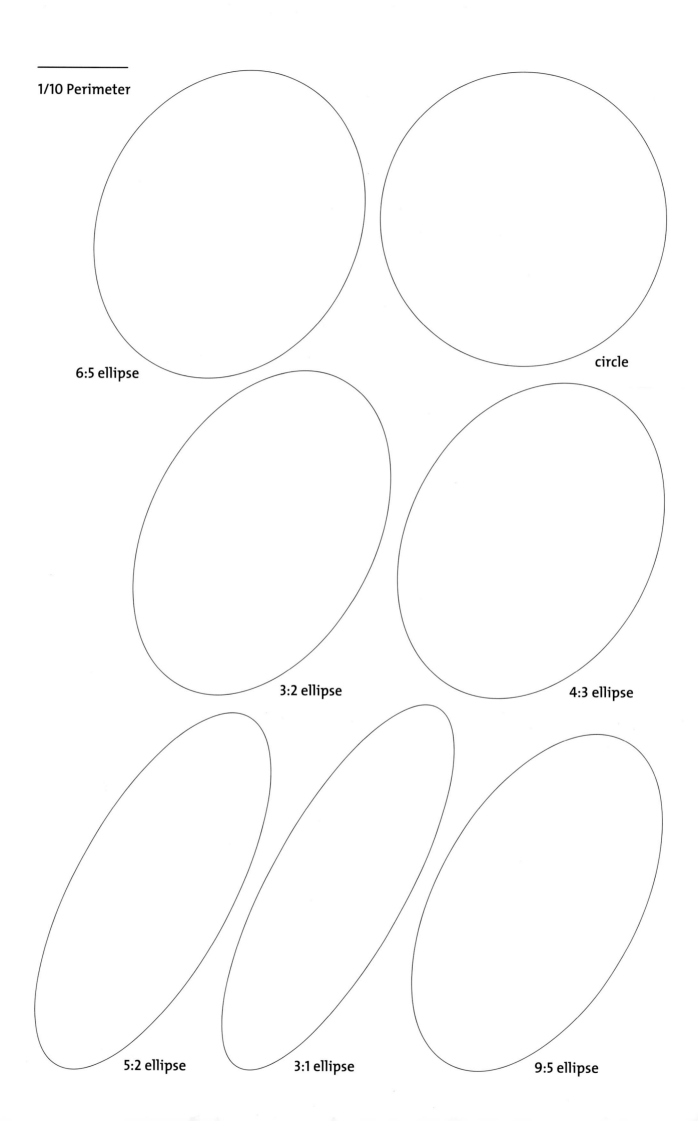

1/10 Perimeter

6:5 ellipse

circle

3:2 ellipse

4:3 ellipse

5:2 ellipse

3:1 ellipse

9:5 ellipse